OUR PLANET EARTH

Oceans

by Karen Latchana Kenney

BLASTOFF! READERS 3

BELLWETHER MEDIA • MINNEAPOLIS, MN

Blastoff! Readers are carefully developed by literacy experts to build reading stamina and move students toward fluency by combining standards-based content with developmentally appropriate text.

Level 1 provides the most support through repetition of high-frequency words, light text, predictable sentence patterns, and strong visual support.

Level 2 offers early readers a bit more challenge through varied sentences, increased text load, and text-supportive special features.

Level 3 advances early-fluent readers toward fluency through increased text load, less reliance on photos, advancing concepts, longer sentences, and more complex special features.

★ **Blastoff! Universe**

Reading Level

Grade K → Grades 1–3 → Grade 4

This edition first published in 2022 by Bellwether Media, Inc.

Library of Congress Cataloging-in-Publication Data

Names: Kenney, Karen Latchana, author.
Title: Oceans / Karen Latchana Kenney.
Description: Minneapolis, MN : Bellwether Media, 2022. | Series: Blastoff! readers. Our planet Earth | Includes bibliographical references and index. | Audience: Ages 5-8 | Audience: Grades 2-3 |
Summary: "Simple text and full-color photography introduce beginning readers to oceans. Developed by literacy experts for students in kindergarten through third grade"-- Provided by publisher.
Identifiers: LCCN 2021011407 (print) | LCCN 2021011408 (ebook) | ISBN 9781644875223 (library binding) | ISBN 9781648344909 (paperback) | ISBN 9781648344305 (ebook)
Subjects: LCSH: Ocean--Juvenile literature.
Classification: LCC GC21.5 .K447 2022 (print) | LCC GC21.5 (ebook) | DDC 551.46--dc23
LC record available at https://lccn.loc.gov/2021011407
LC ebook record available at https://lccn.loc.gov/2021011408

Editor: Rebecca Sabelko Designer: Jeffrey Kollock

Printed in the United States of America, North Mankato, MN.

Table of Contents

What Are Oceans?

Oceans are bodies of saltwater. They cover most of the earth.

Earth has five oceans. They are the Arctic, Atlantic, Indian, Pacific, and Southern. They all connect to make the world ocean!

Oceans have three main zones. The surface zone is bright and full of life.

Pacific Ocean

Famous For

- Largest ocean
- Deepest ocean

Greatest Depth

- More than 36,000 feet (10,973 meters)

Size

- Covers more than 60 million square miles (155 million square kilometers)

Little light reaches the twilight zone, and plants cannot grow. The deep ocean is the largest zone. It is dark and cold.

Oceans affect weather
around the world!
Ocean water stores heat.

Warm water **evaporates** and releases heat into the air. Clouds form in the **atmosphere**. Winds move the heat and clouds around the globe.

Arctic Ocean

Famous For

- Smallest ocean
- Shallowest ocean

Greatest Depth

- 17,850 feet (5,441 meters)

Size

- Covers about 5.4 million square miles (14 million square kilometers)

Oceans have important **cycles**. Along coasts, **tides** move water in and out. **Gravity** from the sun and moon controls tides.

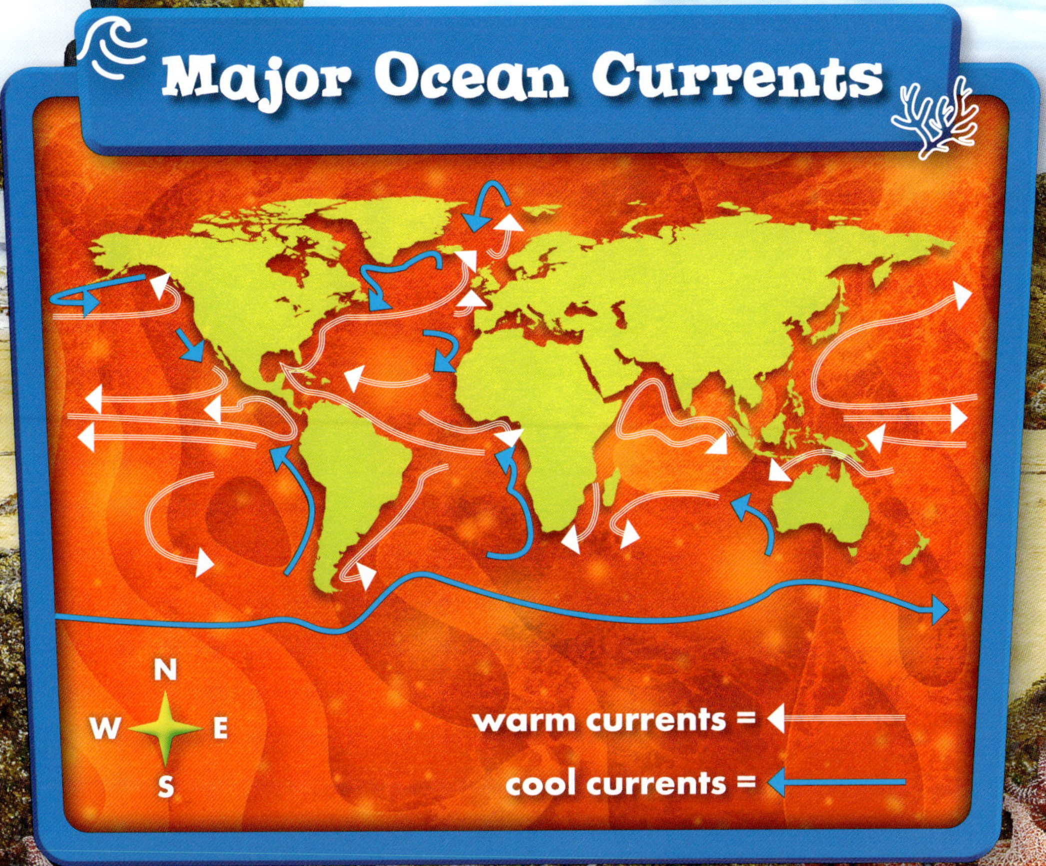

Ocean **currents** move water around the planet. Wind, tides, and water **density** create currents.

Plants and Animals

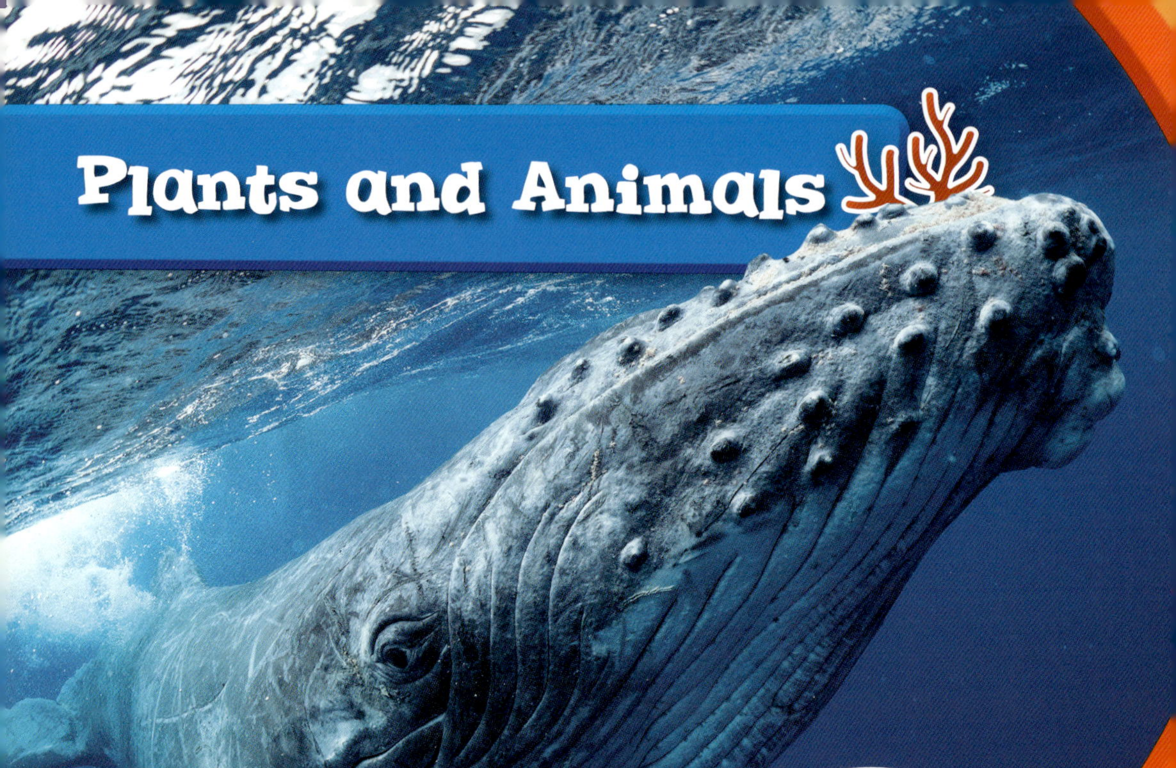

humpback whale

artemia plankton

Oceans have many plants and animals. Along coasts, otters swim through kelp forests. Humpback whales feed on **plankton** and small fish.

Lobsters hunt urchins on **coral reefs**, while sea turtles bite seagrass.

Caribbean spiny lobster

Great white sharks dive for food. Herring speed through the water.

Ocean Animals

crystal jellyfish

Christmas tree worm

In the darkness of the deep ocean, jellyfish and squids make light. Tube worms feed near hot seafloor **vents**.

People and Oceans

People use oceans in many ways. They catch animals for food and sport.

People also take oil and gases from the ocean floor. Big ships carry people and goods across oceans.

Sometimes, human activities harm oceans. Overfishing and loud ships change **habitats** and affect ocean life.

Garbage in ocean waters also damages habitats. **Climate change** makes oceans warmer and harmful to life.

How People Affect Oceans

- **Overfishing changes habitats and harms ocean life**
- **Loud ships make it difficult for some animals to live**
- **Climate change makes oceans warmer and harmful to life**

There are ways to take care of the ocean. Support **sustainable** companies.

Create less waste to keep oceans clean. Everyone can work to keep the oceans and planet healthy!

Glossary

atmosphere—the air surrounding Earth

climate change—a human-caused change in Earth's weather due to warming temperatures

coral reefs—structures made of coral that usually grow in shallow seawater

currents—patterns of water movement

cycles—events that are repeated in the same order

density—how tightly packed something is

evaporates—turns from a liquid into a gas

gravity—the force that pulls objects toward one another

habitats—areas with certain types of plants, animals, and weather

plankton—ocean plants or animals that drift in water; most plankton are tiny.

sustainable—involving practices that do not use up or destroy natural resources

tides—the rise and fall of the oceans that occurs twice a day

vents—openings that allow gas to pass out